the mobile marketer

50 Apps and Tips to up your social game

by
Jenny Holmes

the mobile marketer

50 Apps and Tips to up your social game

By **Jenny Holmes**

Published by:
The Publishing Place
www.ThePubPlace.com

Edited by Caitlin Conley and Ashley Saunders
ISBN # 978-09835095-3-0

Disclaimer: This book is not affiliated with, endorsed by or sponsored by any of the Apps or tools represented in this book. While every effort has been made to introduce these Apps, their benefits, features and costs, the author cannot be responsible for App updates after this book went to print. Readers should review feature detail specific to their smart phone.

Please note that many of these Apps can be set up as new accounts on your computer, then the coordinating App downloaded to your smart phone.

Introduction

Whether you are a marketing professional, public relations guru or entrepreneur, you are in the business of promoting or selling something. It could be a product, a service, a concept or a cause, but staying in the limelight is crucial to your success. New mobile tools make it easy for you to be marketer extraordinaire, get great publicity and create your own limelight. You don't even need an office. In fact, you could go to the beach for several months and nobody would notice you were there (unless of course you post pictures of yourself in your hot new bikini and tag your organization's Facebook page.) This book will teach you how to make the most of all your social media marketing tools and make a way for you to do more mobile marketing from your smart phone. And while there are many Apps listed in this book, there are probably many other similar Apps just as useful. However, in this book, I am sharing my tried and true mobile marketing story. As a marketing director of a large performing arts center with a passion for technology, an unexplainable fondness of my iPhone and the desire to be the best marketing professional ever, this book offers tips, tools, Apps and best practices. But even my story keeps evolving. New Apps are launched daily, so after you get all of these, visit the App store often and subscribe to a monthly App magazine. Technology just keeps getting better and marketing keeps getting easier.

1

Home Base

Your website is the company home base and should showcase all of your social media networking sites with LINKS on the home page. And if those links aren't already there, you need to have your web guy add them immediately. Logos and links for twitter, facebook, your blog, YouTube and any and all social media sites should be easily accessible from the main screen of your home page. Make it easy for people to find you, follow you, be your fan and befriend you.

2

Get Your Own App

You can use a lot of Apps to reach a lot of people but the number one most important thing you can do is take the time to set up your own App. Target marketing doesn't get any better than this! Getting an App used to be expensive and often unaffordable, but all that has changed. If you have some developer skills, you can get an App for the cost of a certificate to Google, $25 and one to Apple, $100 and you'll be available on any smart phone. Otherwise, pay a developer the $300 to $600 fee to get you in this game. Then don't forget to include "download our free App" in all of your marketing efforts. And if you are good at bartering, consider some kind of trade with a developer who wants to get in the App business. We got our App in exchange for 2 tickets to a show at our performing arts center.

3

Facebook

Yes, you probably already have a Facebook fan page for your organization. If you don't, this is the first place to start with your social media efforts. 400 million people are HERE, why go somewhere else? After you set up your page, make sure you make the most out of facebook by posting regularly, putting a pic of whatever you are selling or promoting. Take and post pictures of everything that happens in your organization. You really can never post too much. Add video. Offer free stuff. Have contests. Create events. Post notes. Pay to advertise. There is a whole section under the Advertising tab with tutorials showing you how to set up ads that are more targeted than any other medium out there. And because Facebook is constantly looking for new ways to do things better, they keep making changes and additions, so check in on their advertising section, tutorials and marketing opportunities often to see what great new tools they are offering to advertise in the social media world.

4

Friend or Fan?

If you are a nonprofit, you can have either. However, facebook limits the number of friends to 5,000. Plan for the future and get the Fan page. Otherwise, you'll have a 1,000 people sitting in your friend request queue waiting to be 'friended' and spend weeks trying to move all your friends to your fan page. (Ask me how I know this?) If you already started with a friend page, you can switch to a fan page. Facebook offers tips on how to move everyone over.

5

Ustream

Start your own television show with live streaming TV. It's free and because Ustream thought of everything, you can take your show on the road. First, sign up for an account from your computer. Then download 3 free Ustream Apps on your phone. Ustream Broadcaster lets you broadcast live from your phone. You can advertise Live events, but for those who miss the event, no worries, it archives the video for you. Or maybe you just want to record a video of some event and aren't ready to post just yet. Ustream Recorder App is for just that. Then there is Ustream Viewer. If you are not able to attend an event and your awesome intern is broadcasting live, then you can log onto the event on your phone and watch it. You can also heckle if you want. Ustream offers social interaction and chatting opportunities. And my favorite feature of all: Ustream and Facebook just became best friends. You can now do LIVE chats from within your FAN page. Wow, what a great way to introduce new products, make company-wide announcements, public announcements or show off special guests and events.

6

Facebook Places

Check in. A lot. The more you talk about where you are on behalf of your organization, the more your organization is in the limelight. Wow, what an easy way to check a little box, mention why you are there and tag your friends and coworkers. It shows up on the newsfeed. Easy free publicity.

7

Social voice

It's the human voice, the not-sounding-like-an-I'm-selling-you-something-informercial-voice. It's the conversation you have with your customers. It's the light tone you use to talk about things that are very important to you. It's the passion you exude with your casual words. It's okay to use the word Yep. We hope to see a little wooohooo from you once in a while. You love your job. You love your organization. You love talking about all the stuff that goes on there. Do it in your own voice. The fun social one dude!

8

Interaction

This is the reason for this entire book. Social interaction. No more one way messages. The new marketing invites people to join in and become part of your organization. Social marketing says to your customers, "Hey, we're working on a new project, tell me what you think about this?" Or "how can we do better?" "How can we improve our service or product?" "Which do you like best?" Yes, we are putting ourselves out there and saying, "You, the customer, are now chief planning officers, quality assurance and customer satisfaction managers. Thanks for being just a "like" button or "comment" away from helping us be the best at what we do." And as my pastor would say: it's just the conversation starter for the rest of the week. So get the conversations started. Interaction shows you care about what your customers think and it also opens the door for them to feel more committed to you.

9

Sound Bytes

It's the new way of sending out press releases. 420 characters for Facebook. 140 for Twitter. It's obvious someone has done a study about our attention span, so why not join the crowd and write in bytes. In my 20 years of PR, I have certainly sent my share of REALLY long press releases. But these days if you email releases with the facts, they'll for sure get published in print (shrinking newspaper staffs). And if you have a really interesting topic, a great reporter is going to call you up and say, "Can you send me some photos and can I get an interview? I'd like to do a feature on this topic." Otherwise, the little sound bytes will take on a social media life of their own and get copied, reposted, retweeted and @ and #ed to the masses because they are just the right size.

10

Foursquare

This is just another fun App to let people know where you are while on the go. You can post pictures, share locations and even get discounts and freebies. You can also earn points and badges and become mayor of stuff. More free publicity. And as a professional, I'm certain you'll know when to NOT check in.

11

Twitter

Tweeting is one of the easiest things ever. My jury is still out on the number of followers and the rate our organization accrues them and is it best suited for celebrities. But for something that is so easy to do, why not. Maybe 100 of your Twitter followers aren't the same people that subscribe to your email newsletter or have your App, so it's just one more little way to socialize with some people who took the time to sign up for a free Twitter account and say "I'm going to follow them wherever they go."

12

EZTexting

The number one way people communicate and receive information is via text. You can use a text service one of two ways. If the group of people you are trying to reach is small (few hundred) you can use a free App on your phone and put all the numbers in yourself. If you want to reach the masses, there are several companies out there that offer texting services. People opt in by sending a code word (that you create) to a special text number. You can set up your phone to send out the messages, so if a brilliant marketing idea hits you while you are on the road, just text it. Fees are based on the number of texts you send and the number of people you send them to. For around $50 a month you can hit up about 500 people. Tiered plans let you grow as your text subscribers increase and a website lets you monitor your plan.

13

Dragon Dictation

Sometimes you get a text and need to respond but aren't sure your one finger has time to write out all you need to say. Dictate directly into this App and it creates a text for you. Or maybe, you want to dictate your facebook post or a tweet. Yep, this little App lets you share in a variety of ways including text, email or your social media sites. And if you speak clearly, it is spot on. However, you might read over it before you share, just to make sure. There is a little keyboard if you need to fix it.

14

MailChimp

Your website should have a place for people to sign up to receive your email newsletter. There are a variety of services that let you sign up, use their template, house a database of emails and let you whip out a creative, artsy and interactive piece to those people who opt in to receive it. If you have less than 1,000 friends, it's even free. But the $42 I pay a month to send my bi-monthly news to nearly 5,000 people is a must-have marketing expense. They also have an App, so when you are out and about and someone asks how they can get your newsletter, you can add new subscribers right from your phone in about two seconds. Or if you run into a board member who asks about the open and click rate of your most recent email newsletter, you can whip out your phone and show him the pie chart with the details of that email delivery. Mailchimp cares as much about ROI as you and your board members do.

15

Blog

I use Wordpress. There are a lot of features that allow you to customize your background, add links, videos, photos and most importantly, key search words and excerpts about your blog. And way cooler than that, the RSS feed allows you to link everything together. So, when you write a blog, it will automatically post to your Facebook page and other social media links. Yes, they have an App so you can do it from your phone.

16

Email

I assume you already have an email account set up to go to your phone. If not, this is one of those things to put at the top of your list. And if you have more than one email account, your phone will let you set up to receive more than one account.

17

Camera for photos and videos

If these aren't on your phone, then you might not have the smartest phone of the bunch. Trade up.

18

360

This is a really cool way to take a panoramic photo. You keep moving the camera's viewfinder just a little bit and the App makes it one big pan of a room or a scene not otherwise capturable in a one-shot photo. Not only is it an easy way to get an amazing landscape shot, but you can have a lot of fun with it. Have a person stand at the beginning of the shot, then have them run behind you and get in at the end of the shot. Very Seinfeldish. The photo uploads to facebook completely panoramic. People will notice.

19

iLoader

Yes, you could upload one photo at a time or use iLoader to upload photo batches. This is a great App when you are attending an event or ceremony of some sort where people are getting awards one right after the other. Just keep snapping shots and upload at the end of the event with the push of one button.

20

blurb mobile

If you have some Steven Spielberg aspirations or ever wanted to try your hand at movie making of the story-telling kind, this is the perfect App for you. It lets you combine photos, videos and text to create a mini story. Add text, and even voiceovers to create your own mini-movies, commercials, promos or cover events. There are some time limitations for the free version, but for a couple of dollars you can upgrade to pro. It won't let you make feature-length films but it will help you quickly create and deliver visual messages complete with video and voiceovers.

21

iMovie

So the CEO of your company just gave an incredible welcome presentation and award at an event and you were able to get a great video. You can't wait to post it on your social media sites, but suddenly realize the man standing behind your boss was seriously picking his nose during a short portion of the speech. iMovie is worth the five bucks to be able to go in and clip out that little lapse of judgment, without spending a lot of time in an editing room, and post the fabulous presentation and cash in on a great marketing opportunity.

22

Photoshop Express

Faster than its Adobe full version, this little App lets you crop, add special effects, borders contrast and tints. Not to mention you can crop any unwanted backgrounds or umbrella drinks when you snap a pic of your favorite board member or CEO. Save the new improved version to your phone and upload as usual.

23

Path

Sometimes the events you attend are private. Or maybe large amounts of alcohol are being served at this event. But you are still in charge of photos. So you, as the brilliant marketing person, recognize these photos are not facebook appropriate. Path lets you choose a certain group of friends to share the photos with until such time that you can get them into photoshop and doctor them up to use for future publicity opportunities.

24

Gotta See This

This $1.99 App is a fun way to post pictures in a collage-type way. So let's say you are at an event and there are great shots of people who are supporting your cause or using your product. The App lets you rotate your phone while taking pics. You can even spin in a circle and capture in 360 mode. Then the App creates a collage of a bunch of little snapshots complete with white borders. The final product looks as though you have spent hours creating this collage. It offers a nice finished product in seconds.

25

Grammar App

Whoever/whomever, compliment/complement, over/more than . . . it's easy to get out there and be writing something and suddenly forget the correct word, correct use or spelling of a word. And although your social voice can be light, it still needs to be right. So download one of many free grammar Apps, or pay the $2 - $5 for one that suits your needs.

26

Dictionary.com

You must have this App because AutoCorrect will play with your mind and make you second-guess the spelling of words that you are sure you know. Plus this App has Word of the Day and even sounds it out for you. And there is voice search. You can never know too many words or how to pronounce them.

27

Cramzy

Sometimes you don't have something clever to say, but want to post or tweet to keep the social connection going. There's always a perfect quote. This site features 23,000 of the most profound quotes. Borrow one. Don't forget to give credit.

28

iHandy Translate Pro

You don't have to know a lot of different languages. Just need to know a place to help you convert your clever words to theirs!

29

Google Voice

You can search the Internet while you are driving with the sound of your voice. It's the free Google App. You just say what you are looking for and Google pulls up the website and a phone number. If it doesn't understand you, it prompts you to speak again. And if you still aren't connecting, pull over and type it in. It's just like the google on your big mac.

30

Google Alerts

If you do 500 things to make sure your organization gets noticed, you'll want to know how many times you got noticed. Go into Google and set up Google alerts with your organization's name. You will get an email every time someone prints something about you. This feature alone will save you, your interns or the fee to a clipping service, hours of scouring newspapers to find articles.

31

Voice Memo

Wow, remember when we used to take notes or carry around fancy recorders and have people sit in front of them so we could capture everything they said. Well, if you find yourself being able to get an important impromptu interview or find a happy customer that wants to tell their story, whip this little tool out and record right there. You can email it or text it to someone as soon as you finish recording. It saves as a Quicktime file, which you can later use in videos or podcasts. It's also a great productivity tool if you need to record a to-do reminder for yourself for the next time you are in your office (if you ever go back there).

32

The News

I have Apps for USA Today, New York Times and NPR all on my phone. I take what I call a little news bath every morning and make sure I start my day knowing what's up in the rest of the world and in my industry. And then because I am nosey, I read my Facebook newsfeed to see what all of my friends are up to. You can never know too much.

33

Seth Godin

Seth Godin is one of the most insightful and brilliant marketing gurus, author of numerous books, creator and inventor and the consummate mentor to entrepreneurs and marketing professionals across the world. His unique perspective and writings, both books and blogs, make me stop and think about how I go about the business of marketing each day and why I make the choices I make and thoughts on how I might look at things through the eyes of my customer. You must download his free App and you must read his blog on a daily basis.

Whatever your industry, there is an App for that. And because I am very focused on being the best marketing and PR professional I can be, I am going to continue to find Apps that teach me new things and provide me with best practices in my industry. While this book speaks to those interested in marketing and publicity, there are thousands of Apps out there and many geared toward specific industries. Search by 'category' in the App store.

34

LogMeIn

Put the App on your phone, your iPad or whatever device you could readily use to access the computer in your office. There will be large files, graphic design work, fliers and other items that you just don't carry around on your phone. But you can easily access that big MAC in the office remotely with this handy little tool. It's well worth the $29.99 price tag to be able to access everything that is just too big to be mobile.

35

iCloud

It's the latest greatest way you store everything somewhere else then access from any device. Upload your music, photos, Apps, calendars and documents from your computer. Then access them from your iPad, your iPhone, iPod and whatever device you own. First 5 GB are free, after that, there are some nominal fees. At the time this went to print, it wasn't actually available yet, but get it on your hit list because it's coming and you need it.

36

Mouse Pad

So you've logged in to your office computer from your iPad and realize you need to make a little change in a brochure you designed at the office. Since designing with one finger can be tricky, turning your phone into a mouse pad will make you feel like you are sitting right there at your desk doing the work.

37

Google Analytics

ROI. ROI. ROI. Knowing the return on your investment is the most important component of marketing. If you are going to spend money or time on marketing and publicity, this way or that way, you need to know if it is working. Google analytics is another free tool that lets you log in and find out how many people visited your website, how many pages they looked at, how long they stayed, what country they are from and just a variety of statistics to keep you informed. The graphics are great. The reports are easy to read and can easily be copied into your monthly report so everyone in your organization knows how much you know about your customers and more importantly potential customers.

38

Google Goggles

Google keeps adding amazing features and this is one of my favorites yet. So let's say you are out covering an event and mingling with some potential new customers. You don't know much about their company and want to have intelligent conversations. You can slip your camera out and snap a picture of their company logo and Google Goggles will go find the company and pull up their website – because it can literally recognize the photo of the logo. Amazing technology. It works on landmarks, books, artwork, labels, logos and much more.

39

PR Newswire

You probably have a local media list that you send information to, but some of your stories, new product launches or new services might be of interest to a bigger market. And this is how you let people know. PR Newswire is THE place to distribute news releases and get reports on the distribution. You can sign up online. It's free if you are a nonprofit. It's an annual membership of $195 if you are a for-profit. Then you can choose from a variety of plans to launch a press release. PR Newswire has already gathered the important contacts from all over the world and lets you choose geographically and demographically where you'd like to send that release. You pay a fee per release based on the reach you choose. The free App for your phone also lets you GET all the news. If journalists and newsrooms all over the world rely on this service, it's probably something you can count on too.

40

MarketingProf

This amazing App gives you daily marketing tips, offers free webinars and conferences and all things to keep providing you with the tools you need to stay up-to-date in the marketing world. And even though you may have a degree in marketing or PR, the rules are constantly changing and this little App will remind you of things you might have forgotten and teach you the new stuff.

41

iTunes U

It's an online university. iTunes U lets you access college lectures without the pricey tuition and early morning class hours. You can get lectures on everything from social marketing, to starting your own business to conducting business in other cultures. This is just another way to stay on top of your business game.

42

iTunes

Have you thought about podcasting? It's as simple as conducting an interview, recording the audio portion of your Ustream TV show or giving your own lecture. You can record from the little Voice Memo tool on your phone – or use a microphone and headset and record from your computer. You then save it as an MP3 audio file, submit it to Apple so they can approve. Then you invite people to download your free podcasts on iTunes.

43

Become an eReporter

Chances are, there was a day when you sent out press releases to announce your events and prayed the TV stations would find the events newsworthy enough to show up with a camera and put you on the 6 p.m. news. Those days are over. Now, you just get your own video clips and email to them. Most local news stations have created social interactive opportunities they call eReporting. They want their viewers to send photos and videos of what's going on in their communities. They may or may not use them, but they invite people to submit. Wow, isn't this a dream come true for marketing and PR professionals. Our goal is to stay in the limelight and get all the publicity we can for our organization. So, call your local news stations and offer to be an eReporter for your organization. Make sure you ask their preferred format and make it easy for them to use your clips. My stations recommend a horizontal video of at least twenty seconds, no more than thirty seconds long. I email the video right from my phone. Then email details of the event. I'm on the news all the freaking time!

44

TurboScan

It's like having a little copy machine in your pocket. Use this App to scan important documents from your phone. You can create PDFs, jpegs and put multiple scans into one file.

45

ePrint

So now that you have scanned that important document and need to print it, free ePrint App lets you do that right from your phone in one of two ways. You can set up your office printer with an IP address, so the document will print directly to your office and be there next time you show up. Or, it will let you search for a neighbor's wireless printer, who might be nice enough to let you borrow their ink. And sometimes, as marketer extraordinaire, you'll not only snap the perfect shot to post on Facebook or send out as tweet, but you'll want to print it. Yep, you can take and print a photo right from within the ePrint App. However, this feature only connects you to certain HP printers. Buy one now or hang tight because other ePrinters are going to follow.

46

Free Wi-Fi Finder

Let's say your significant other surprised you with the gift of an iPad but failed to get the 3GS. So now when you travel, you have to look for Internet hotspots so you can connect. This free Wi-Fi finder works anywhere in the world from your phone's GPS to help you locate free Wi-Fi Internet without driving around until you find it on your own.

47

My Marketing Plan

Yes, you still need a plan. It's so easy to get caught up in the fast-paced world of RSS feed and how quickly you can text, tweet and fire off a message, that you might forget the importance of having a plan of action. At just under $25, there is an App for that too. The App works with all IOS versions and walks you through step by step how to write the plan on your phone.

48

Groupme

There are times when you need to make your circle smaller and share information, photos, your location and even conference calls with just a select group of office mates, committee members or your management group. This perfect little App lets you connect in all these ways for FREE.

49

Thinkfree office mobile

There will be times when you need to access a file that is on your computer (back in the place you call 'office'). There is no point in carrying everything around on your phone. Just access them with this App anytime you want. Photos, PDFs, Word, Excel, Powerpoints—all easily accessible. Save the files to the free thinkfree.com and then connect to them from your phone anytime you want.

50

iPhone tips and tricks App

This App is like a mini book featuring important tips complete with screenshots. As a first generation iPhone user, I thought I knew everything there was to know about my phone. But this App teaches you how to get HOME faster, how to multi-task better and even offers troubleshooting tips for any and everything that can go wrong with your phone. If there is an App for an iPhone, there is probably a similar App for all other smart phones, so do the research.

Made in the USA
Monee, IL
04 November 2026

67382375R00059